YOUR KNOWLEDGE HAS VALUE

- We will publish your bachelor's and master's thesis, essays and papers

- Your own eBook and book - sold worldwide in all relevant shops

- Earn money with each sale

Upload your text at www.GRIN.com and publish for free

Bibliographic information published by the German National Library:

The German National Library lists this publication in the National Bibliography; detailed bibliographic data are available on the Internet at http://dnb.dnb.de .

Imprint:

Copyright © 2014 GRIN Verlag, Open Publishing GmbH
Print and binding: Books on Demand GmbH, Norderstedt Germany
ISBN: 9783668534490

This book at GRIN:

http://www.grin.com/en/e-book/375278/shsearch-a-method-for-fast-remote-homology-detection

Mohamed Baddar, Noha Yousri

SHsearch. A Method for Fast Remote Homology Detection

GRIN Publishing

GRIN - Your knowledge has value

Since its foundation in 1998, GRIN has specialized in publishing academic texts by students, college teachers and other academics as e-book and printed book. The website www.grin.com is an ideal platform for presenting term papers, final papers, scientific essays, dissertations and specialist books.

Visit us on the internet:

http://www.grin.com/

http://www.facebook.com/grincom

http://www.twitter.com/grin_com

SHsearch: a method for fast remote homology detection

M. Baddar

Contents:

SHsearch: a method for fast remote homology detection

M. Baddar

Abstract—Remote homology detection is the problem of detecting homology in cases of low sequence similarity. It is a hard computational problem with no approach that works well in all cases. Methods based on profile hidden Markov models (HMM) often exhibit relatively higher sensitivity for detecting remote homologies than commonly used approaches. However, calculating similarity scores in profile HMM methods is computationally intensive as they use dynamic programming algorithms. In this paper we introduce SHsearch: a new method for remote protein homology detection. Our method is implemented as a modification of HHsearch: a remote protein homology detection method based on comparing two profile HMMs. The motivation for modification was to reduce the run time of HHsearch significantly with minimal sensitivity loss. SHsearch focuses on comparing the important submodels of the query and database HMMs instead of comparing the complete models. Hence, SHsearch achieves a significant speedup over HHsearch with minimal loss in sensitivity. On SCOP 1.63, SHsearch achieved 88X speedup with 8.2% loss in sensitivity with respect to HHsearch at error rate of 10%, which deemed to be an acceptable tradeoff.

Index Terms—biological sequence classification, hidden Markov models

———————————————————

1 INTRODUCTION

Analysis of large scale sequence data has become an important task in computational biology and comparative genomics, inspired in part by numerous scientific and technological applications such as the biomedical literature analysis or the analysis of biological sequences. Protein homology detection has attracted particular interest due to its critical role in many biological applications. One of these applications is protein function prediction [1] which is an important task in drug design [2] .The process of drug design can be divided into two phases. First phase is searching for a target protein whose molecular function is to be moderated, in many cases blocked, by a drug molecule binding to it. Second phase is selecting a suitable drug that binds to the protein tightly, is easy to synthesize, is bio-accessible and has no adverse effects such as toxicity. The knowledge of protein function can be of significant help in both phases. However, protein annotation with functional information lags behind in the rapidly increasing amount of sequence data resulting from the numerous ongoing genome sequencing projects. Consecutively, the manual analysis and annotation of protein function via laboratory experimental procedures, which is a low throughput process, became no longer effective. The availability of a large amount of protein sequences data and the vitality of the problem motivated the development of high throughput computational methods for protein function prediction.

A family of state-of-the-art protein function prediction approaches relies on gathering information about protein functions from different sources [1]. These sources of information include protein homology [3] [4], gene expression analysis [5], protein interaction networks analysis [6], phylogenic trees and profiles analysis [7], and literature text mining [8]. One of the most effective means of inferring the function of a newly sequenced protein is to detect what functions are performed by homologous proteins [4]. Two proteins are homologous if they share a common ancestor. Since the actual sequence of the common ancestor is unavailable, sequence homology can only be inferred by statistical means.

Dynamic programming algorithms, such as the Needleman-Wunsch [9] and Smith-Waterman algorithms [10], or related heuristic algorithms, such as BLAST [11] and FASTA [12], can be used to assign to each sequence in the database a score indicating the likelihood that this sequence is homologous to the query based on their similarity score. However, these approaches have low sensitivity in remote homology detection [13]. Consequently, other approaches have been developed targeting the detection of remote homologies. For example, motif based approaches [14] are developed based on comparing query and database sequences' motifs. Other approaches works by building a profile for a query sequence which is constructed from a multiple alignment of sequences closely related to the query [15]. Then this profile is scored against all database sequences. A significant advance is achieved by comparing query profiles to database profiles [16] [17] [18] [19]. Database profiles are built from the multiple alignments of clusters of closely related sequences in a database. This type of profile-profile comparison approaches gives relatively higher sensitivity in detecting remote homologies than other methods in the literature [19].

1

The rest of the paper is organized as follows. In section 2 we review background information and related work. In section 3 we propose our method. In section 4 we illustrate the experimental setup and performance measures. In section 5 we show and discuss results. In Section 6 we present our conclusion.

2 BACKGROUNDS AND RELATED WORK:

In this section, the background of homology detection methods is reviewed. We illustrate the evolution of different methods for solving different issues related to homology detection.

2.1 BASIC HOMOLOGY DETECTION METHODS

Over the past decades, various methods have been proposed for the task of homology detection. The key idea behind most of these methods is to calculate a score to measure the similarity between a given query and all databases sequences. The query and database sequences are classified as homologous if the score exceeds some predefined threshold.

The most commonly used score for comparing two sequences is the Smith-Waterman score [10]. It is calculated as the score of the local optimal pairwise alignment of two sequences. For retrieving sequences similar to a newly sequenced protein (a query), Smith-Waterman algorithm [10], BLAST [11] and FASTA [12] are the most commonly used methods. These methods use the Smith-Waterman score to measure the similarity between sequences in different ways. BLAST and FASTA are heuristic algorithms which use certain assumptions and approximations. Both programs first identify very short exact matches between the query and database sequences. Next, the best short hits from the first step are extended to look for longer stretches of similarity. Finally, the best hits are optimized with some form of dynamic programming. On the other hand, the Smith-Waterman approach is a completely dynamic programming tool which effectively makes all possible pairwise comparisons to all of the database sequences. Hence, it is a much more sensitive technique as compared to BLAST and FASTA, but it is much more computationally expensive and slower than any of them. However, all mentioned methods' sensitivity degrades significantly in remote homology detection as they depend only on pairwise similarity score between the query and database sequence [13].

2.2 Remote homology detection methods

For the remote protein homology detection task, numerous methods have been developed to achieve relatively higher sensitivity than the methods mention in Sec. 2.1. For example, the method proposed in [14] is based on the presence of sequence motifs. The motif content of a pair of sequences is used to define a similarity that is used as a kernel for a classifier. Motifs represent limited, highly conserved regions of proteins [20]. By focusing on comparing motifs, important clues to a protein's function can often be revealed even if it is not globally similar to any known protein in databases. Hence, this approach works well in remote homology detection as it alleviates depending on low global pairwise similarity score and focus on conserved segments which usually have relatively higher similarity in distantly related sequence.

An improvement in remote homology detection is achieved by developing methods based on profile-sequence comparison [15]. A profile is usually built from query sequence(s) either from the multiple alignment of the family of closely related sequences to the query or directly from an input multiple alignment. The profile allows one to distinguish between conserved positions that are important for defining members of this family and non-conserved positions that are variable among the members of the family. Furthermore, it describes exactly what variation in amino acids is possible at each position by recording the probability for the occurrence of each amino acid along the multiple alignment. Therefore, a profile contains more information about the sequence's family of closely related sequences than a single sequence. Hence, profile-sequence methods provide higher sensitivity for detecting remote homologies than BLAST, FASTA and similar methods based on comparing query and database sequences.

A typical example of profile-sequence methods is PSI-BLAST [21]. Given a query sequence, PSI-BLAST first performs a quick BLAST search to retrieve database sequences closely related to the query. Then a multiple

2

alignment is created from this initial search's hits. After that, a Position Specific-Scoring Matrix (PSSM) is generated from this alignment. This PSSM is used as a profile which is scored against all database sequences and a new PSSM is created by combining search results to original PSSM. The last step is repeated until no significant new search results are retrieved. A PSI-BLAST variant; Cascade PSI-BLAST [22]detects remote homologies by performing cascade propagation of PSI-BLAST search results through all hits identified for a query. This approach allows effective detection of remote homologies for that query sequence.

2.3 Profile hidden Markov model procedures for remote homology detection

A significant breakthrough in profile-sequence methods is achieved by using a hidden Markov model (HMM) as a profile for modeling a family of related sequences instead of a PSSM or other types of standard profiles [23] [24]. SAM [25] and HMMer [26] [27] are the most popular software packages that use HMMs as profiles. Krogh introduced an HMM architecture well suited for representing a profile for multiple sequence alignments [28]. For each consensus column in the multiple alignment, a 'match' state models the distribution of residues allowed in this column. An 'insert' state and 'delete' state at each column allow for insertion of one or more residues between this column and the next, or for deleting the consensus residue. An emission probability distribution is associated with each match and insert state to model the residues emitted from that state. Also, a transition probability is assigned to transit form a state to another based on the model architecture.

A profile HMM can be considered as an abstraction model which by following its states' emission and transition probabilities, a set of sequences is generated that are closely related to the sequences used to build the model. Figure 1a shows a sample multiple sequence alignment, and a profile HMM built form this alignment is shown in figure 1b.

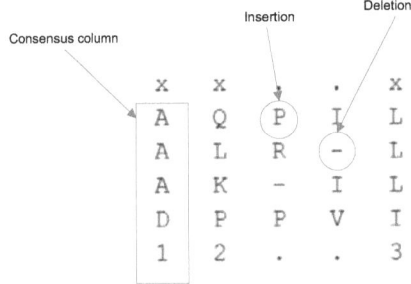

(a)   Protein multiple sequence alignment

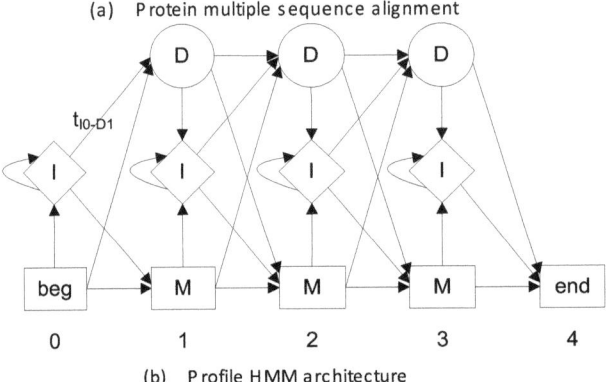

(b)   Profile HMM architecture

Fig. 1. The Profile HMM in (b) is built from the multiple sequence alignment shown in (a) with three consensus columns. These three consensus columns are modeled by three match states (rectangles: M1-M3), each has 20-residue emission probabilities vectors. Insertions in the multiple alignment are modeled by insert states (diamonds: I0-I3) which also have 20-residue emission probabilities vectors each. Delete states (circles: D1-D3) are silent states with no emission probabilities. Transitions are represented by arrows with a transition probability assigned for each arrow, for example t_{I0-D1}. The model length is measured by the number of match states; hence this model is of length 3.

A profile HMM has several advantages over a PSSM and other types of standard profiles [29]. It has a formal probabilistic basis and a consistent theory behind match, insertion and deletion scores, in contrast to other types of profiles which use heuristic methods. Also, a profile HMM applies a statistical method (Henikoff and Henikoff [30]) to estimate the true frequency of a residue at a given position in the alignment from its observed frequency. These estimated frequencies are used to build match and insert states' emission probabilities vectors. On the other hand a PSSM uses the observed frequency itself to assign an emission probability for that residue .This means that a profile HMM derived from a relatively small number of sequences can be of equivalent quality to a PSSM created from a larger number of aligned sequences [20]. Hence a profile HMM provides better modeling capability and homology detection sensitivity.

However, the methods relying on profile HMMs are computationally intensive as dynamic programming algorithms are used for building a HMM and scoring its similarity against database sequences. These algorithms called Forward-Backward (for building model) and Viterbi (for scoring a model against a sequence) have a worst case time complexity of $O(NM^2L_{seq})$, where N is the number of sequences , M is the number of HMM states and L_{seq} is sequence length [23]. Another approach is used to build a profile HMM from a multiple alignment called Maximum A-Posteriori algorithm (MAP) [31] is of time complexity $O(NL_{align}^2)$ where L_{align} is the number of columns in the multiple alignment.

Many variants to HMMer have been proposed to either increase computational speed or sensitivity. One variant is proposed in [32] which extracts submodels from a query profile HMM, called sub-HMMs. These sub-HMMs represent parts of the HMM that model the most conserved parts (motifs) of the sequences used to build the original HMM. Sub-HMMs are shorter, information rich models with the same architecture of a profile HMM. Based on that definition, extracting sub-HMMs works as follows. First the Kull-back-Leibler divergence (KL-divergence or relative entropy) [33] is calculated for each mach state, then a series of normalization and smoothing steps is performed and the most information rich HMM regions are excised from the original profile HMMs. KL-divergence is calculated as in (1)

$$h_i = \sum_j \ln\left(\frac{E_{M_i}(j)}{B_j}\right) E_{M_i}(j)$$

(1)

where h_i is the KL-divergence value of match state M_i , $E_{M_i}(j)$ is the estimated emission probability of residue j in the emission distribution vector E of M_i . B_j is the background distribution of residue j. KL-divergence measures the amount of information carried by this state's emission distribution relative to the background distribution. The result of these extraction steps is an ordered set of extracted sub-HMMs. These extracted submodels are then scored against each database sequence. This set of submodels is ordered from left to right based on their position in the original model. Figure 2 shows sample sub-HMMs and sample emission vectors.

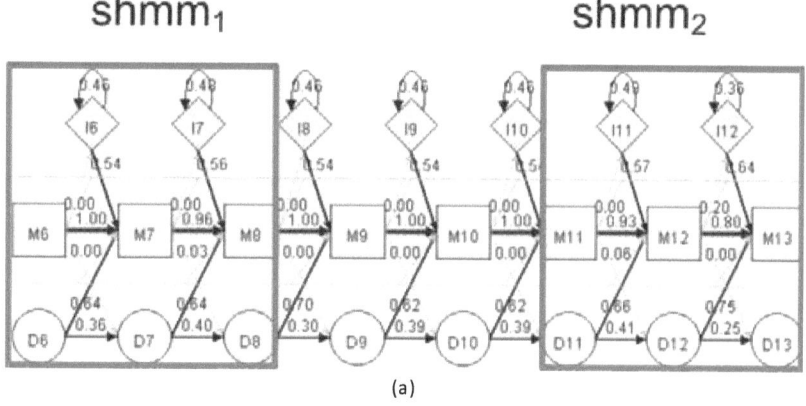

(a)

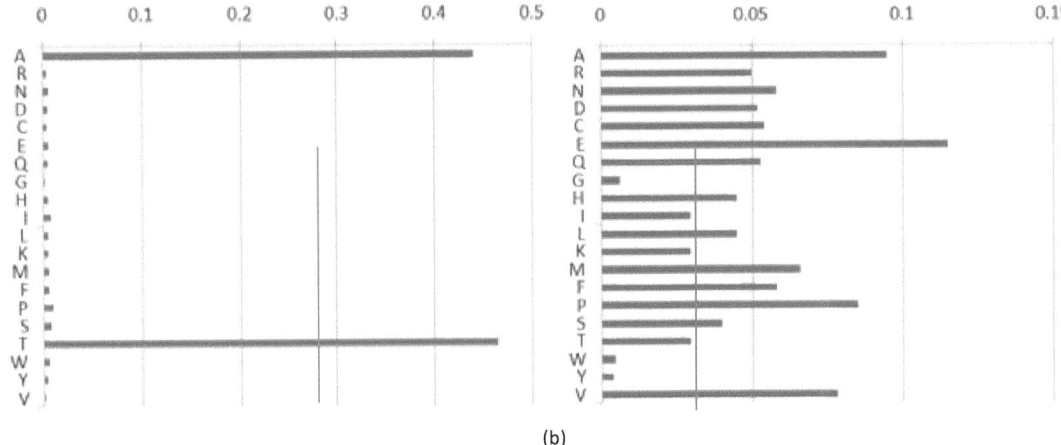

(b)

Fig. 2. (a) Two sample sub-HMMs ($shmm_j; j = 1,2$) extracted from a profile HMM with match states belonging to that sub model highlighted in red. (b) Sample protein residues emission distributions for two mach states M_6(the distribution on the left) and M_9(the distribution on the right). M_6 has high KL-divergence value hence it belongs to a sub-HMM. On the other hand M_9 has low KL-divergence value so it doesn't belong to a sub-HMM.

The target of this sub-HMM based homology detection approach is to extend the use of profile-HMMs to be used in highly localized sequence similarity searches that focus on shorter conserved parts of sequence rather than entire domains or global similarities. Another variant to HMMer is HMMERHEAD [34] which filters a query profile to the most important parts for homology detection. HMMERHEAD initially generates and identify significant "words". This step consists of identifying ungapped four residue words from a profile HMM's match state emission vectors that possess a probability above some threshold. A word score is calculated as the sum of the log-odds emission probabilities of the word's residues, as determined from the profile-HMM. These words are then identified in the database sequences using a deterministic finite automaton. After that, each word identified in a database sequence is the seed for an ungapped alignment between the sequence and the profile-HMM. By focusing on important short words, HMMERHEAD achieves a 20X speed over HMMer with 4% loss in sensitivity.

A major advance in profile based remote homology detection methods was achieved by comparing profiles to profiles. Profile-profile comparison approaches can be considered as an extension to profile-sequence approaches which leverages profiles' advantages in similarity search for both database and query sequences. Several programs for homology detection have recently been developed based on this idea: PROF_SIM [16], COMPASS [18], LAMA [17] and HHsearch [19]. These programs were shown to be significantly more sensitive than PSI-BLAST and have been applied for identifying evolutionary links between protein families previously thought to be unrelated [17] [18] [35]. HHsearch is based on comparing two profile HMMs by calculating the co-emission probability (P_{coem}) [36] for these two models. Co-emission probability is the probability that two HMMs independently generate the same sequence, that is for models HMM_i and HMM_j generating sequences over and alphabet Σ we compute P_{coem} according to (2)

$$P_{coem}(HMM_i, HMM_j) = \sum_{s \in \Sigma^*} P_{HMM_i}(s) P_{HMM_j}(s) \qquad (2)$$

where $P_{HMM_i}(s)$ is the probability that HMM_i generates sequence s. A detailed dynamic programming algorithm for computing P_{coem} is introduced in [36]. P_{coem} expresses how two HMMs are similar which in turn reflects the similarity of the groups of sequences represented by these two HMMs [25]. By combining the advantages of profile-profile methods and profile-HMMs, HHsearch achieves relatively higher sensitivity in detecting remote homologies than other homology detection methods. In the following section we present how we modify HHsearch for achieving higher computational speed with minimal loss in sensitivity.

3   Proposed Method

The experimental results in [19] show that HHsearch's sensitivity for detecting remote homologies is relatively higher than other homology detection methods in the literature. However, HHsearch has a high computational complexity and takes significant run time since it uses a dynamic programming algorithm similar to the Smith-Waterman algorithm for scoring the similarity of two HMMs. The high computational complexity of HHsearch was the main motivation for developing our method.

The main idea behind the modification is focusing on scoring short "key" submodels of database profile HMMs against submodels extracted from the query profile HMM, instead of scoring two complete models. These key submodels are the most important submodels for homology detection, as they carry most of the information needed for classifying database profile HMMs as homologous or non-homologous to a given query. The work we introduce in this paper is similar to HMMERHEAD [34](see Sec. 2.2) in an effort to reduce search time significantly with minimal loss in sensitivity.

Since the time complexity for scoring two HMMs depends on the lengths of these models (will be discussed in Sec. 3.3), our modification significantly reduces the run time of HHsearch with a minimal loss in sensitivity. Also by focusing on selecting the most important submodels to use in comparison for homology detection, the sensitivity loss becomes minimal as shown in results in Sec. 4. In section 3.1 we start with rules and definitions. We describe the details of SHsearch method in section 3.2.

3.1 Rules and definitions

In this section we present rules and definitions used in SHsearch method. In rule 1 we present conditions for aligning two sets of submodels extracted from two HMMs. In definition 1 we introduce S_2 : a new score for measuring the similarity of two profile HMMs. In definition 2 we introduce a novel relevancy score that is used for extracting key submodels. We present a formal definition for hierarchal clustering in databases of sequences and homology and non-homology clustering levels in definition 3

Rule 1 (submodels alignment): Let HMM_i and HMM_j be two different profile hidden Markov models, l_i and l_j be the lengths (number of match states)of these models. Let $P_{coem}(HMM_i, HMM_j)$ be the co-emission probability of the two HMMs as described in [36]. A similarity score of the two HMMs based on P_{coem} introduced in [36] and is calculated as shown as in (3)

$$S_1(HMM_i, HMM_j) = \frac{P_{coem}(HMM_i, HMM_j)}{\sqrt{P_{coem}(HMM_i, HMM_j) * P_{coem}(HMM_j, HMM_j)}} \quad (3)$$

S_1 is normalized i.e. $0 \leq S_1 \leq 1$. Recall from Sec. 2 that sub-HMM is a shorter model with the same architecture and parameters set of a profile HMM. Hence, S_1 can be used to score pairs of sub-HMMs and to construct an alignment for the two sets of extracted submodels. Let $shmm_{ia}$ be a sub model extracted from HMM_i with index a, then the alignment rules are as follows:

1. Every submodel in the first set is aligned with a submodel in the other set iff both have the highest S_1 score amongst all possible pairings. This rule can be stated formally as follows:
   $shmm_{ia'}$ is aligned with $shmm_{jb'}$ iff $S_1(shmm_{ia'}, shmm_{jb'}) \geq S_1(shmm_{ia}, shmm_{jb}) \forall a, b \, ; 1 \leq a \leq n_i \, , 1 \leq b \leq n_j$, where n_i, n_j is the number of extracted submodels from HMM_i and HMM_j .

2. No cross alignment is allowed, that is:
   For any submodels: $shmm_{ia'}$, $shmm_{ia''}$, $shmm_{jb'}$ and $shmm_{jb''}$ If $shmm_{ia'}$ is aligned with $shmm_{jb'}$ and $a'' > a'$,then $shmm_{ia''}$ can only be aligned with $shmm_{jb''}$,iff $b'' > b'$

Figure 3a shows an alignment of two sets of submodels that satisfies the two conditions, while the alignment in figure 3b has a cross alignment and hence violates condition 2.

6

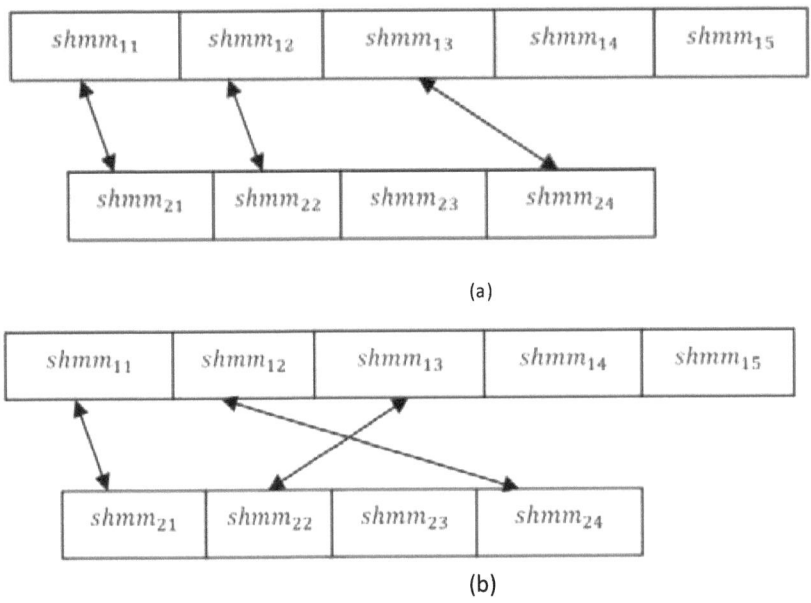

Fig. 3. Alignments for two sets of extracted submodes (a) consistency conditions are satisfied. (b) Cross alignment violates condition 2

This alignment rule is a modification of the rule presented in [37]or aligning sequences based on their conserved segments.

Definition 1 (Weighed HMMs similarity score): Let HMM_i and HMM_j be two different hidden Markov models, l_i and l_j be the lengths of these models, $(shmm_{ia^*}, shmm_{jb^*})$ be the set of aligned submodels based on rule 1 and n_p be the number of aligned pairs of sub models. We define S_2 as in (4)

$$S_2(HMM_i, HMM_j) = \frac{\sum_{p=1}^{n_p}(l_{ia^*} + l_{jb^*})S_1(shmm_{ia^*}, shmm_{jb^*})}{\sum_{p=1}^{n_p}(l_{ia^*} + l_{ib^*})} \tag{4}$$

where $S_1(shmm_{ia}, shmm_{jb})$ is the similarity score mentioned in rule 1, l_{ia} and l_{jb} are lengths of $shmm_{ia}$ and $shmm_{jb}$, where $\sum_{a=1}^{n_i} l_{ia} < l_i$ and $\sum_{b=1}^{n_j} l_{jb} < l_j$, n_p is the number of pairs of aligned submodels where $n_p \leq min(n_i, n_j)$ n_i, n_j are the number of extracted submodels from HMM_i and HMM_j, respectively. S_2 is also normalized i.e. $0 \leq S_2 \leq 1$.

The logic behind weighting S_1 by lengths of submodels is that longer submodels represent longer conserved segments of sequences, which are more important in identifying related families of sequences than shorter segments. S_2 is used to score the similarity between a query and database HMMs as we will detail in section 3.2. Hence SHsearch achieves a speedup over HHsearch ,as S_2 calculations are limited to shorter key submodels, instead for calculating similarity of the complete HMMs. Figure 4 shows an example for calculating S_2 for two sample HMMs; HMM_1 and HMM_2 and their extracted and aligned set of submodels.

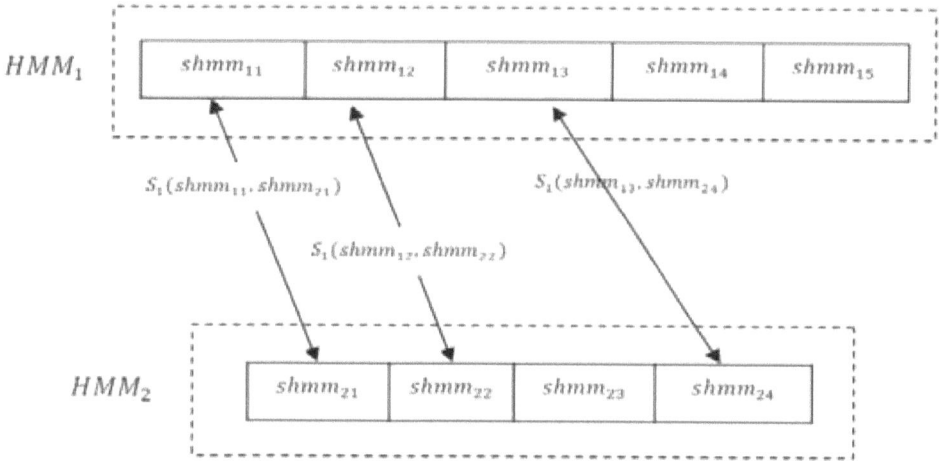

Fig. 4. An example of $S_2(HMM_1, HMM_2)$ calculation. Arrows represent aligned pairs of submodels. Number of extracted submodels: $n_1 = 5$, $n_2 = 4$, and number of aligned pairs of submodels $n_p = 3$. S_1 is calculated for each aligned pairs. Hence S_2 is calculated as follows:

$$S_2(HMM_1, HMM_2) = \frac{(l_{11}+l_{21})S_1(shmm_{11},shmm_{21})+(l_{12}+l_{22})S_1(shmm_{12},shmm_{22})+(l_{13}+l_{24})S_1(shmm_{13},shmm_{24})}{(l_{11}+l_{21})+(l_{12}+l_{22})+(l_{13}+l_{24})}$$

Definition 2 (submodels relevancy score): Given a database that provides a hierarchical clustering for its sequences (as will be shown in definition 3), let HMM_k be a profile HMM built from a multiple alignment of a set of sequences of a cluster at level-1 C_{1k} (See definition 3). Let $shmm_{ka}$ be a submodel extracted from HMM_k with index a, as shown in Sec. 2.3. We define a relevancy score for each submodel as in (5)

$$RS(shmm_{ka}) = \frac{\sum_{i'} S_1(shmm_{ka}, shmm_{k'a'})}{\sum_{i''} S_1(shmm_{ka}, shmm_{k''a''})} \quad a \neq a' \neq a'' \text{ and } k \neq k' \neq k'' \tag{5}$$

where $shmm_{k'a'}$ represents all submodels in the homology cluster of the original model HMM_k (See definition 3). On the contrary, $shmm_{k''a''}$ represents all submodels in all non-homology clusters with respect to HMM_k. This score measures for each submodel how much it is close to other homologous submodels and how much it is distant relative to non-homologous submodels. The higher value of RS means that this submodel is more important for homology detection. The submodels with the highest RS value are selected as "key" submodels.

The derivation of this relevancy score is inspired by the term frequency-inverse document frequency (TF-IDF) numerical score [38]. This score reflects how important a word is to a document in a collection or corpus and is often used as a weighting factor in information retrieval and text mining. In our case a keyword corresponds to a key submodel and a document collection corresponds to a homology cluster (See definition 3).

Definition 3 (Homology and non-homology clustering levels): Given a set of sequences $S_1, S_2, ..., S_{N_s}$ where N_s is number of sequences in a database, a hierarchical clustering of sequences can be constructed as shown in figure 5. Assume that there are H levels in the clustering hierarchy with level-1 being the lowest level and level H being the highest. As cluster levels increase, the sequences become less related (less similar) to each other. Let C_{xk} be a cluster at level x with k where $1 \leq x \leq H, k \geq 1$. Each level in the hierarchy represents a certain degree of similarity and evolutionary relationship between sequences belonging to every cluster at that level.

A profile HMM (HMM_k) is built from the sequences of each level-1 cluster C_{1k}. The Smith-Waterman pairwise similarity score and information about functional and structural similarity can be used as a pairwise similarity measure for constructing the hierarchical clusters. Based on the hierarchical clustering scheme we present, we define both a homology and a non-homology clustering levels; x_h and x_{nh} as follows:

(i) *Let $C_{x_h k}$ be any cluster at level x_h with index k and S_i, S_j be any two sequences belonging to $C_{x_h k}$, where $1 \leq x_h \leq H$ and $i, j, k \geq 1$. We call x_h a "homology clustering level" iff S_i and S_j are homologous for all values of i*

8

and j. In other words; any two sequences belonging to the same cluster at level x_h are classified as homologous. $C_{x_h k}$ is called a homology cluster for all profile HMMs belonging to it.

(ii) *Let $C_{x_{nh} k}$ and $C_{x_{nh} k'}$ be any two different clusters at level x_{nh} with indices k and k' where $1 \leq x_{nh} \leq H$, $k \neq k'$ and $k, k' \geq 1$. Let S_i be any sequence belonging to $C_{x_{nh} k}$, $S_{i'}$ be any sequence belonging to $C_{x_{nh} k'}$. We call x_{nh} a "non-homology clustering level" iff S_i and $S_{i'}$ are non-homologous for all values of i, i', k, k'. In other words; any two sequences belonging to two different clusters at level x_{nh} are classified as non-homologous. $C_{x_{nh} k}$ is called a non-homology cluster for all profile HMMs not belonging to it. For any hierarchical clustering of database sequences $x_{nh} > x_h$.*

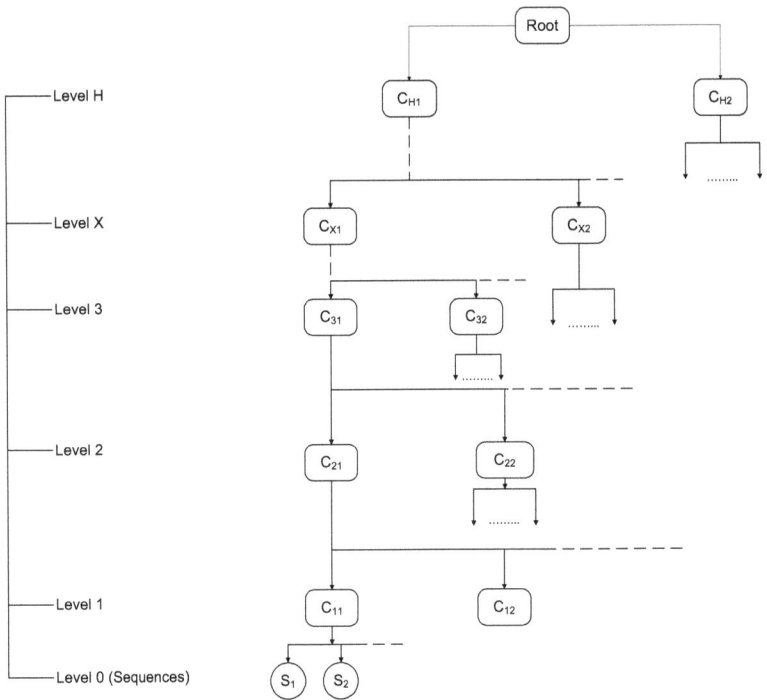

Fig. 5. Hierarchical clustering of database sequences $S_1, S_2, \ldots, S_{N_s}$. C_{xk} is a cluster at level x ; $1 \leq x \leq H$ with index $k \geq 1$. The Smith-Waterman pairwise similarity score and information about functional and structural similarity can be used along with other sources of information to measure similarity between sequences and to construct clusters.

3.2 SHsearch method

In this section we describe the details of SHsearch method. The main idea behind the proposed method is limiting calculations of scoring query and database profile HMM to the shorter "key" submodels in the database. These key submodels are obtained by extracting submodels from the database HMMs then filtering them to the ones with the highest relevancy score (See definition 2 in Sec. 3.1). The input database must have a hierarchical clustering for its sequences with homology and non-homology clustering levels defined as shown in definition 3. This hierarchical clustering can be predefined like SCOP database [39] (as will be detailed in Sec.4) or we can use evolutionally relationship information and similarity score values between sequences to construct hierarchical clusters. SHsearch works in two main phases:

1. Database preprocessing phase
    1.1. A multiple alignment is built for sequences in each level-1 cluster (C_{1k}) in the hierarchy using ClustalW if there is no existing alignment created in the database.

1.2. A profile HMM (HMM_k) is built for each alignment using the Maximum A-Priori (MAP) algorithm.

1.3. For each HMM_k, a set of submodels ($shmm_{kl}$) is extracted using the method explained in Sec. 3.

1.4. For each $shmm_{kl}$ the relevancy score $RS(shmm_{kl})$ is calculated. Then the set of submodels is sorted in descending order based on the calculated RS values. After that, these submodels are filtered to the subset of the highest RS score based on a given filtering ratio $0 \leq r_{filter} \leq 1$ as will be detailed in preprocessing algorithm (figure 7) this subset is called "key" submodels.

2. Search phase:

2.1. Given an input sequence s, perform a BLAST search to get sequences closely related to s and then a multiple alignment is built from results.

2.2. A query HMM ($qHMM$) is built from the alignment built in step 2.1 using the MAP algorithm. Then all submodels are extracted from this query HMM.

2.3. The proposed weighted similarity score (S_2) is calculated between the query profile ($qHMM$) and each database HMM (HMM_k). If $S_2(qHMM, HMM_k) \geq t_{homology}$, where $t_{homology}$ is a predefined homology threshold, $0 \leq t_{homology} \leq 1$, then $qHMM$ and HMM_k are classified as homologus. Therefore, all sequences used to build HMM_k are also classified as homologous to query sequence s.

A flowchart of SHsearch method is shown in figure 6. The details of preprocessing and search phases are shown in the algorithms in figures 7 and 8.

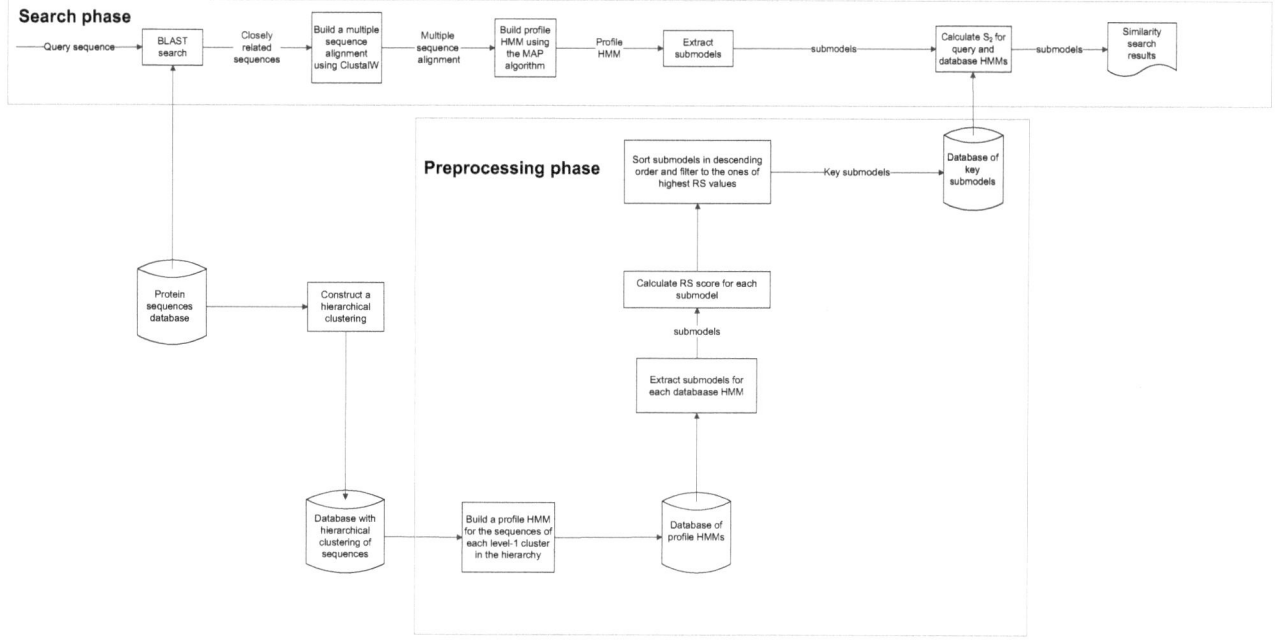

Fig.6. SHsearch method flowchart

Algorithm 1 Preprocess	
1.	**INPUT**
2.	$DBhierCluster$: a database of sequences with hierarchical clustering
3.	r_{filter}: a filtering ratio for extracting key submodels
4.	**OUTPUT**
5.	$keySubHMMsDB$: a database of key submodels
6.	**PROCEDURE** shsearch_preprocess ($DBhierCluster$)
7.	$\{C_{1k}\}$: be the set of level-1 clusters in $DBhierCluster$
8.	$hmmsDB$: a database of profile hidden Markov model
9.	hmm_k : a hidden Markov model built from sequences in C_{1k}
10.	$shmms_k$: a set of sub-HMMs extracted from hmm_k
11.	$shmm_{kl}$: a sub-HMM extracted from hmm$_k$ with index l
12.	/*build profile hidden Markov model using the MAP algorithm*/
13.	**foreach** C_{1k}
14.	/*build a multiple sequence alignment*/
15.	msa_k = ClustalW(sequences(C_{1k}))
16.	hmm_k=MAP_build(msa_k)
17.	add (hmm_k, $hmmsDB$)
18.	end **foreach**
19.	/*extract key submodels from each profile HMM*/
20.	**foreach** $hmm_k \in hmmsDB$
21.	/*extract submodels for each database hidden Markov model*/
22.	$shmms_k$=extract_subHMMs(hmm_k)
23.	N_{shmms_k}= number of submodels in $shmms_k$
24.	/*calculate the relevancy score for each extracted submodel as shown in definition 2 in Sec. 4.1*/
25.	**foreach** $shmm_{kl} \in shmms_k$
26.	RS_{kl}=calculate_RS ($shmm_{kl}$)
27.	/*filter key submodels*/
28.	/* sort submodels based on RS value*/
29.	$sorted\_shmms_k$ =decreasing_sort ($shmms_k$ by RS_{kl})
30.	end **foreach**
31.	/*calculate number of submodels to be extracted*/
32.	$N_{filter,k}$=ceiling($r_{filter} * N_{shmms_k}$)
33.	/* create list of key submodels */
34.	**for** (i=1 to $N_{filter,k}$)
35.	add($sorted\_shmms_k[i]$,$keySubHMMs_k$)
36.	/*add a reference to hmm_k and its key submodels to the data base of key submodels*/
37.	add (hmm_k,$keySubHMMs_k$, $keySubHMMsDB$)
38.	end **for**
39.	end **foreach**
40.	**return** $keySubHMMsDB$
41.	**END PROCEDURE**

Fig. 7. Preprocessing algorithm for SHsearch process

Algorithm 2 Search
1.   **INPUT**
2.   *Qseq*: query sequence
3.   *SeqsDB* : database of sequences
4.   *keySubHMMsDB*: database of key submodels (the output of preprocess phase)
5. $t_{homology}$: homology threshold
6.   **OUTPUT**
7.   *homologousSeqsList* : Set of sequences homologous to Qseq
8.   **PROCEDURE** shsearch_search($Qseq, SeqsDB, keySubHMMsDB, t_{homology}$)
9.   *seedSeqs*: be the set of seed sequences closely related to the query to build a profile
10. msa_Q : the multiple sequence alignment built from seed sequences similar to the query
11.   *qHMM*: the query profile hidden Markov model
12. $shmms_Q$: the set of submodels extracted from the query HMM
13. $keySubHMMs_k$: the set of key submodels for hmm_k
14.   /*retrieve set of closely related sequences to the query*/
15.   *seedSeqs*= BLAST_search($Qseq, SeqsDB$)
16.   /*build a multiple sequence alignment*/
17. msa_Q=ClustalW(*seedSeqs*)
18.   /*build a profile hidden Markov model*/
19.   *qHMM*=MAP_build(msa_Q)
20.   /*extract submodels from the query profile HMMs*/
21. $shmms_Q$=extractSubHMMs(*qHMM*)
22.   /*search for homologies*/
23.   **foreach** hmm_k
24. $keySubHMMs_k$=getKeySubModels($keySubHMMs_k$,k)
25. S_2=calculate_S_2($hmm_k, qHMM$)/* calculation is performed using extracted submodels*/
26.   **if** ($S_2 \geq t_{homology}$)
27.   add (sequences(C_{1k}), *homologousSeqsList*)
28.   **end if**
29.   **end foreach**
30.   **return** *homologousSeqsList*
31. **END PROCEDURE**

Fig. 8. Search algorithm for SHsearch method

3.3 Time complexity analysis:

As shown in section 4.2 the main computation operation in HHsearch and SHsearch is scoring the similarity of two HMMs. The similarity of two HMMs (or sub-HMMs) is based on calculating the co-emission (P_{coem}) of these two models as mentioned in section 3.1. The main difference of the two algorithms is that HHsearch calculates the similarity of two complete models where SHsearch combines the similarity score of shorter submodels; this is the key reason for the speed gain. The algorithm used for calculating this co-emission probability introduced in [36] is similar to the Smith-Waterman local alignment algorithm [10] . The time complexity for calculating the Smith-Waterman score (SW-score) for two sequences seq_1 and seq_2 [40]is calculated as in (6)

$$Time\ complexity\ of\ calculating\ SW\text{-}score(seq_1, seq_2)$$
$$= 2 * O(l_{seq_1}l_{seq_2}) + O(l_{seq_1} + l_{seq_2}) \tag{6}$$

where l_{seq_1} and l_{seq_2} are the lengths of seq_1 and seq_2 respectively. Similarly, the time complexity of calculating the co-emission probability is as shown in (7)

12

$$Time\ complexity\ of\ calculating\ P_{coem}(HMM_i, HMM_j)$$
$$= 2 * O(l_i l_j) + O(l_i + l_j) \tag{7}$$

where l_i and l_j are the lengths(number of match states) of HMM_i and HMM_j. Accordingly the time complexity of comparing two HMMs in SHsearch can be calculated as in (8)

$$Time\ complexity\ of\ calculating\ S_2$$
$$= \sum_{p=1}^{n_p} (2 * O(l_{ia^*} l_{jb^*}) + O(l_{ia^*} + l_{jb^*})) \tag{8}$$

where l_{ia^*} and l_{ib^*} are the lengths of aligned sub models $shmm_{ia^*}, shmm_{jb^*}$. To compare time complexities in (7) and (9) we can approximate l_{ia^*} and l_{jb^*} to their average values l_{i-avg} and l_{j-avg}. Based on this approximation, the average time complexity of calculating S_2 is as in (9)

$$Approximate\ average\ Time\ complexity\ of\ calculating\ S_2$$
$$= 2 * O(n_p l_{i---avg} l_{j-avg}) + O(n_p l_{i-avg} + n_p l_{j-avg}) \tag{9}$$

Given that $n_p \leq min(n_i, n_j)$ and typically $n_p l_{i-avg} < l_i$ and $n_p l_{j-avg} < l_j$, and by comparing (7) and (9) we can see that the calculation time of S_2 time is on average less than P_{coem} and therefore SHsearch is faster than HHsearch.

4 EXPERIMENTAL EVALUATION

We study the performance of our method in terms of sensitivity and run time for detecting remote homologies using standard benchmark datasets for protein sequence analysis.

4.1 Datasets and experimental setup

We test our proposed method on the task of remote protein homology detection. This task tests the ability to build a classifier that would correctly detect proteins remotely homologous to a newly sequenced protein. We construct a benchmark dataset based on SCOP 1.63 database [41] to test the proposed method's performance on remote homology detection. SCOP (Structural Classification of Proteins) [39] database aims to categorize proteins into structural hierarchy of classes, folds, superfamilies, families and domains as shown in figure 9. SCOP is used by many previous works (e.g. [42] [43] [44]).

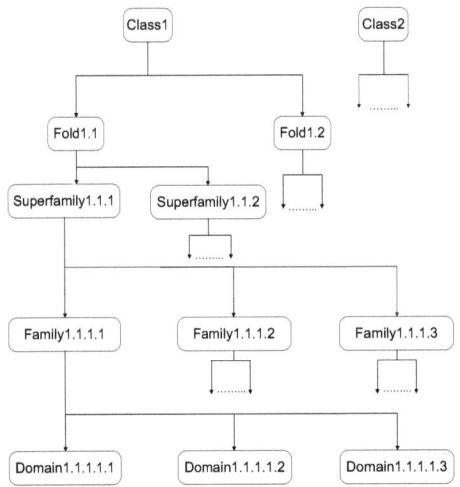

The steps for constructing the remote homology dataset are similar to what is introduced in [19] as follows:

1. The sequences of the SCOP database are filtered to a maximum sequence identity of 20% ('SCOP-20') which are obtained from the ASTRAL server [45]. We obtain a set of "seed" sequences each corresponds to a single domain.

2. An alignment is built from each seed sequence by PSI-BLAST with up to eight iterations. An inclusion threshold of 10^{-4} is the last iteration and 10^{-5} in previous iterations is used. A set of "synthetic" domains is built around these sequences.

3. A multiple sequence alignment is build using ClustalW for the sequences of each synthetic domain.

4. A profile HMM is build form each alignment using the MAP algorithm.

After applying these 4 steps on SCOP database we obtain a remote homology datasets of profile HMMs for distantly related sequences. This constructed dataset then will be used to compare the performance of our proposed remote homology method against other methods. The performance measures we will discuss are sensitivity, run time and scalability.

4.2 Run time analysis

We test SHsearch and HHsearch1 (the basic version of HHsearch) using each sequence in the constructed dataset as a query (All-against-All search). The parameter r_{filter} affects the number of database key submodels used for homology detection and hence the number of submodels to be aligned. As shown in (8) the time complexity of calculating S_2 is proportional to number of aligned submodels n_p, thus r_{filter} affects the run time of SHsearch. The parameter $t_{homology}$ doesn't affect run time of SHsearch. Hence, for run time analysis we vary r_{filter} values only. The measures we use for to compare the run time of SHsearch and HHsearch1 is the average HMM-HMM comparison time for both SHsearch and HHsearch1. To compare the run time of both SHsearch and HHsearch1 we calculate the speed up achieved by SHsearch over HHsearch1 as in (10)

$$speedup = \frac{Average\ computation\ time\ for\ HMM\text{----}HMM\ comparison\ HHsearch1}{Average\ computation\ time\ for\ HMM-HMM\ comparison\ SHsearch} \tag{10}$$

4.3 Sensitivity analysis

Homology detection methods classify each database sequence as homologous to a query if their similarity exceeds a predefined threshold. Hence, sensitivity [46] is used to measure the performance of homology detection methods. Sensitivity is defined as in (11)

$$Sensitivity = \frac{number\ of\ true\ positives}{number\ of\ true\ postives + number\ of\ false\ negatives} \tag{11}$$

where true positives (TP) are the correctly identified homologous pairs, and false negatives (FN) are the homologous pairs that aren't detected. Following SCOP, we classify any two domains (and their corresponding sequences and HMMs) as homologous if they are members of the same superfamily. Domains from different

14

classes are classified as non-homologous. All other pairs are considered as 'unknown' in the benchmark as their evolutionary relationship can't be ascertained [19].

The sum of the numbers of true positives and false negatives is the total number of homologous pairs. Hence, in SCOP the total number of homologous pairs is the total number of all possible pairs in belonging to the same superfamily. The number of true positives is the number of pairs classified as homologous and belonging to the same superfamily. The number of false positives is the number of pairs classified as homologous and belonging to different classes. Also we calculate the error rate value as shown in (12)

$$error\ rate = \frac{number\ of\ false\ positives}{number\ of\ true\ postives + number\ of\ false\ negatives} \tag{12}$$

where false positives (FP) are the non-homologous sequences pairs identified as homologous. We run an All-against-All search for SHsearch and vary the values of r_{filter} and $t_{homology}$ to measure SHsearch's sensitivity. Also we perform All-against-All runs for HHsearch1, HMMer, PSI-BLAST and BLAST and compare their sensitivity to the values obtained from testing SHsearch.

4.4 Scalability Analysis

We test the scalability of SHsearch by measuring how its speedup values scales as the database size grows .We construct smaller databases by randomly selecting a proportion of the sequences in the constructed remote homology dataset. Then we test SHsearch with different r_{filter} values on these generated databases and measure the method's speedup for different databases' sizes.

4.5 Parameters selection

The parameter r_{filter} affects run time as it controls the number of submodels used for homology detection, hence it controls the time complexity of comparing two HMMs. Also, r_{filter} affects sensitivity as it controls the amount of information lost due to selecting some of the submodels. On the other hand, $t_{homology}$ affects sensitivity only; as $t_{homology}$ increases, the homology detection classifier becomes more "restrictive". For run time and sensitivity analysis, the range of r_{filter} values is set to be between 0.5 and 0.9 with an increment of 0.1, and the value of $t_{homology}$ is fixed at 0.8. By using this range of values for r_{filter}, we obtain significant speed gains with acceptable sensitivity losses. For sensitivity analysis, the $t_{homology}$ values' range is set to be from 0.95 to 0.25 with a decrement of 0.05. This range results in different values for true positives and false positives and the sensitivity curves shown in figure 10 are traced out. In scalability analysis we use values of $r_{filter} = 0.5, 0.9$, $t_{homology} = 0.8$ and we select of a proportion of 50, 60,70,80,90 and 100% of the sequences in the remote homology dataset.

5 Results and Discussion

In this section we discuss the results obtained for each type of analysis mentioned in Sec. 5.

5.1 Run time analysis results

By varying r_{filter} values and fixing the value of $t_{homolgy} = 0.8$ we obtain different run time values of HMM-HMM comparison as shown in table 1.

Table 1: SHsearch speedup over HHsearch1

15

Homology detection Method	Average HMM-HMM comparison time(μs)	Speedup
HHsearch1	8126	1
SHsearch($r_{filter} = 0.9$)	92	88.6
SHsearch($r_{filter} = 0.8$)	75	108.2
SHsearch($r_{filter} = 0.7$)	67	122.4
SHsearch($r_{filter} = 0.6$)	65	126
SHsearch($r_{filter} = 0.5$)	62	131.7

In our experiment, the average of profile HMM's lengths in the constructed dataset is found to be 1054. By substituting these values for l_i and l_j in (7), we find the average time complexity of calculating P_{coem} for scoring the similarity of two complete HMMs is 2225204 basic operations (where basic operations are arithmetic summation and multiplication). On the other hand, in SHsearch at $r_{filter} = 0.9$ the average value for the number of aligned pairs of sub-HMMs (n_p) are found to be 12 , and the average values of the lengths of the submodels (l_{i-avg} and l_{j-avg}) is 30.5; hence by substituting in (9) the average time complexity for calculating S_2 is 23058 basic operations. According to these calculations the expected speedup is $\frac{2225204}{23058} = 96.5$. The measured experimental speedup at these parameters is 88.6 as shown in table 1. The experimental speedup value is less than the expected analytical value as there exists and initial overhead in SHsearch for extracting query submodels.

5.2 Sensitivity analysis results

By run HHsearch1 and SHsearch with r_{filter} and $t_{homology}$ with the ranges mentioned in Sec. 5.4 we obtain the sensitivity analysis curves as shown in figure 10. The error rate is the factor that controls the ratio between the number of true positives and false positives and can be represented as a straight line as shown in figure 9 where dashed line represents an error rate of 10% and dotted line represents an error rate of 15%. In our experiment on SCOP 1.63 database there is total number of true positives of 41510 and false positives of 1.8x10⁷.

As we see in figure 10, it is observed that despite that SHsearch's sensitivity is less than HHsearch, it still outperforms HMMer3, PSI-BLAST and BLAST. Also from figure 10 and the speedup results shown in table 1 we can see that SHsearch achieves a significant speedup over HHsearch1 with a minimal sensitivity loss. The reason behind the minimal sensitivity loss is that the information lost due to filtering HMM to a subset of its

16

extracted submodels is minimized by selecting the most important submodels for the homology detection task.

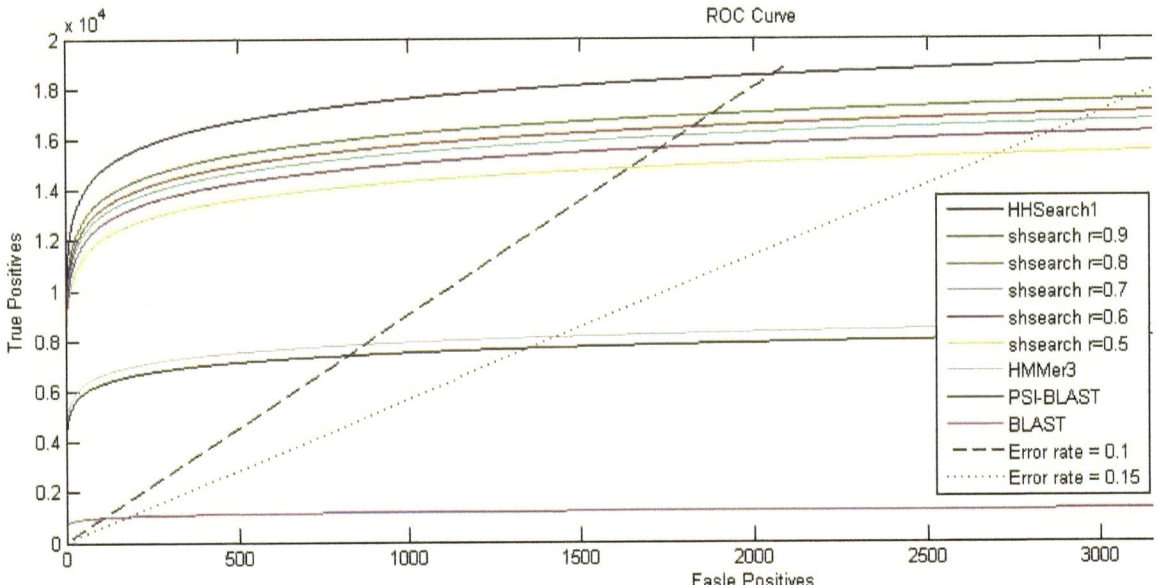

Fig. 10. Sensitivity analysis of SHsearch at different values of r_{filter} against different homology detection methods .The higher the curve, the higher the sensitivity of the corresponding method. We observe that as r_{filter} increases, the sensitivity curve of SHsearch rises up. Also we can see that the sensitivity curves of SHsearch at different values of r_{filter} is slightly lower than the curve of HHsearch, which means that SHsearch's sensitivity is slightly less than HHsearch. Furthermore, the curves of SHsearch are significantly higher than the curves of other homology detection methods.

5.3 Speed-up vs. Sensitivity Analysis

Table 2 and 3 show the speed gain against sensitivity values and sensitivity loss of SHsearch relative to HHsearch1 at error rate of 10% and 15%. We use these values for error rate, as we can see in figure 10, the straight lines intersect the curves at points where the search methods' curves "saturate", i.e. they detect all the true positives they can. The values in these two tables are calculated from the intersection of the error rate lines with the sensitivity analysis curves.HHSearch1 finds 18520 true positive out of 41510 (found via experimentation), so its sensitivity at error rate 10% = $\frac{18520}{41510}$=44.6%. The relative sensitivity loss shown in the tables is calculated as in (13)

$$\begin{aligned} &Relative\ sensitiviy\ loss\ for\ SHsearch\ relative\ to\ HHsearch1 \\ &= \frac{number\ of\ true\ positives\ in\ HHsearch1 - number\ of\ true\ positives\ in\ SHsearch}{number\ of\ true\ positives\ in\ HHsearch1} \end{aligned} \quad (13)$$

Table 2
Sensitivity and Speedup values for SHsearch and HHsearch at error rate of 10%

Homology detection Method	Sensitivity	Speedup	Relative sensitivity Loss
HHsearch1	44.6%	1	0%
SHsearch($r_{filter} = 0.9$)	40.8%	8.8	8.2%
SHsearch($r_{filter} = 0.8$)	39.7%	9.3	11.3%
SHsearch($r_{filter} = 0.7$)	38.7%	9.8	13.3%
SHsearch($r_{filter} = 0.6$)	37.6%	11.2	15.8%
SHsearch($r_{filter} = 0.5$)	35.7%	13.3	20.04%

Table 3
Sensitivity and Speedup values for SHsearch and HHsearch at error rate of 15%

Homology detection Method	Sensitivity	Speedup	Relative sensitivity Loss
HHsearch1	46.2%	1	0%
SHsearch($r_{filter} = 0.9$)	42.2%	8.8	8.65%
SHsearch($r_{filter} = 0.8$)	40.98%	9.3	11.3%
SHsearch($r_{filter} = 0.7$)	40.11%	9.8	13.2%
SHsearch($r_{filter} = 0.6$)	38.95%	11.2	15.69%
SHsearch($r_{filter} = 0.5$)	36.9%	13.3	20.13%

From tables 2 and 3 we can see that for SHsearch, as speedup increases, sensitivity decreases. The reason is that as r_{filter} decreases, the number of submodels in homology detection decreases, hence the time complexity of comparing time decreases and speedup occurs. However, the decrease in number of used submodels increases the information loss which leads to decrease in sensitivity. In situations where we need high sensitivity, we recommend using r_{filter}=0.9 and $t_{homology} \geq 0.8$, this will achieve a speedup of 8.8X over HHsearch and sensitivity $\geq$ 42%. In other situation where we want a quick search to get initial homology results we can use lower values for r_{filter}.

5.4 Scalability analysis

We test SHsearch against different sizes databases that are constructed for the remote homology database as illustrated in section 5.3. Tables 4 and 5 and figures 11 and 12 shows the scalability analysis results by listing measured values of speedup against different databases sizes at two values for r_{filter} and a fixed value for $t_{homology}$.

Table 4
Scalability analysis results for SHsearch at $r_{filter} = 0.5$ and $t_{homology} = 0.8$

Database size (in number of domains)	Speedup
1846	31.6
2214	50.1
2584	64.5
2953	109.3
3322	122.5
3691	131.7

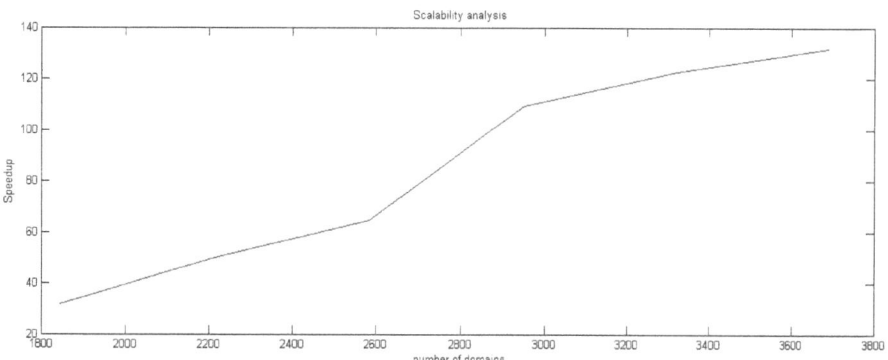

Fig. 11. Scalability analysis curve at $r_{filter} = 0.5$
and $t_{homology} = 0.8$ (Plotting of table 4 results)

Table 5
Scalability analysis results for SHsearch at $r_{filter} = 0.9$ and $t_{homology}$=0.8

Database size (in number of domains)	Speedup
1846	14.8
2214	27.5
2584	39.9
2953	55.4
3322	65.5
3691	88.6

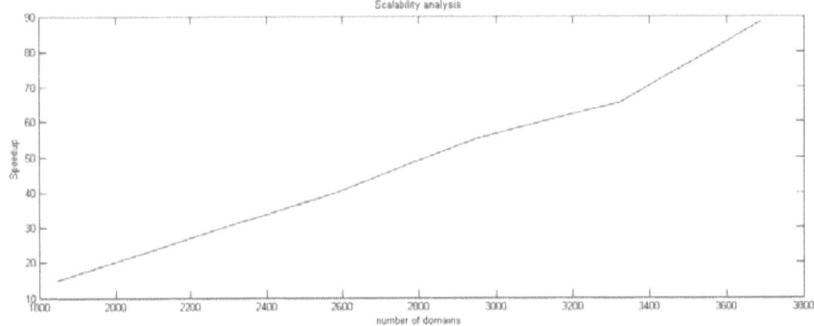

Fig. 12. Scalability analysis curve at $r_{filter} = 0.9$
and $t_{homology} = 0.8$ (Plotting of table 5 results)

From the results in figure 11 and 12 we can infer that SHsearch has approximately a linear scalability for different databases' sizes. In Figure 11, we can see that SHsearch achieves a 10X speedup for each increase of 250 domains in database size. SHsearch has a one-time overhead for building query extracting submodels form the query, which doesn't exist in HHsearch. This overhead effect becomes less when the database size increases. This linear scalability makes SHsearch more attractive to use other than HHsearch for very large databases.

6 Conclusion

We presented SHsearch; a new remote homology detection method. SHsearch is developed as a modification for HHsearch, a remote homology detection method based on comparing two profile-HMMs. SHsearch focuses on comparing "the most informative" submodels extracted from the query and database HMMs, that's the reason of minimal loss in sensitivity. Also, the aggregate lengths of these submodels is significantly less than the original HMM, hence SHsearch achieves a noticeable speedup over HHsearch. We have benchmarked SHsearch against HHsearch using a remote homology dataset constructed based on SCOP database. SHsearch achieves 88X speedup with 8.2% loss in sensitivity at 10% error rate which deemed to be an acceptable tradeoff. In addition, SHsearch has a relatively higher sensitivity than HMMer3, PSI-BLAST and BLAST. Furthermore, SHsearch exhibits almost linear scalability for speedup over HHsearch as database size increases. This makes SHsearch attractive to use especially in larger size database.

7 References

[1] Gaurav Pandey,Vipin Kumar and Michael Steinbach, "Computational Approaches for Protein Function Prediction," Department of Computer Science and Engineering, University of Minnesota, Twin Cities, TR 06-028, 2006.

[2] T. Lengauer and R. Zimmer, "Protein structure prediction methods for drug design," *Briefings in Bioinformatics*, vol. 1, no. 3, pp. 275-288, September 2000.

[3] Yaniv Loewenstein, Domenico Raimondo, Oliver C Redfern,James Watson,Dmitrij Frishman,Michal Linial, Christine Orengo, Janet Thornton, and Anna Tramontano, "Protein function annotation by homology-based inference," *Genome Biology*, vol. 10, no. 2, p. 277, February 2009.

[4] William Noble Grundy, "Family-based Homology Detection via Pairwise Sequence Comparison," in *Second Annual International Conference on Computational Molecular Biology*, 1998, pp. 94-100.

[5] M.G. Walker, W. Volkmuth, E. Sprinzak D. Hodgson and T. Klinger, "Prediction of gene function by genome-scale expression analysis: prostate cancer-associated genes.," *Genome Research*, vol. 9, no. 12, pp. 1198-1203, 1999.

[6] A. Vazquez, A. Flammini, A. Maritan and A. Vespignani, "Global protein function prediction from protein-protein interaction networks," *Nature Biotechnology*, vol. 21, no. 1, pp. 697-700, May 2003.

[7] "Protein function predictions based on the phylogenetic profile method," *Critical Reviews in Biotechnology*, vol. 28, no. 4, pp. 233-238, 2008.

[8] R. Nair and B. Rost, "Annotating protein function through lexical analysis.," *AI Magazine*, vol. 25, no. 1, pp. 45-65, 2004.

[9] Saul B. Needleman and Christian D. Wunsch, "A general method applicable to the search for similarities in the amino acid sequence of two proteins," *Journal of Molecular Biology*, vol. 48, no. 3, pp. 443-453, March 1970.

[10] T.F. Smith and M.S. Waterman, "Identification of common molecular subsequences," *Journal of Molecular Biology*, vol. 147, no. 1, p. 195, March 1985.

[11] S.F. Altschul , W. Gish, W. Miller, E.W. Myers and D.J. Lipman, "Basic local alignment search tool," *Journal of Molecular Biology*, vol. 215, no. 3, pp. 403-410, October 1990.

[12] W.R. Pearson and D.J. Lipman, "Improved tools for biological sequence comparison," *National Academy of Sciences of the USA*, vol. 85, no. 8, pp. 2444-2448, April 1988.

[13] B. Rost, "Twilight Zone of Protein Sequence Alignments," *Protein Engineering Design and Selection*, vol. 12, no. 2, pp. 85-94, February 1999.

[14] Asa Ben-Hur and Douglas Brutlag, "Remote homology detection: a motif based approach," *Bioinformatics*, vol. 19, no. Supplement 1, pp. i26-i33, 2003.

[15] Micheal Gribskov, Andrew D. Mclachlant, and David Eisenberg, "Profile analysis: Detection of distantly related proteins," *Proceeding of National Academy of Sciences of USA*, vol. 84, no. 1, pp. 4355-4358, July 1987.

[16] G. Yona and M. Levitt, "Within the twilight zone: a sensitive profile-profile comparison tool based on information theory," *Journal of Molecular Biology*, vol. 315, no. 5, pp. 1257-1275, February 2002.

[17] S. Pietrokovski, "Searching databases of conserved sequence regions by aligning protein multiple-alignments," *Nucleic Acids Research*, vol. 24, no. 19, pp. 3836-3845, October 1996.

[18] R. Sadreyev and N. Grishin, "COMPASS: a tool for comparison of multiple protein alignments with assessment of statistical significance," *Journal of Molecular Biology*, vol. 326, no. 1, pp. 317-336, February 2003.

[19] J. Söding, "Protein homology detection by HMM-HMM comparison," *Bioinformatics*, vol. 21, no. 7, pp. 951-960, April 2005.

[20] C.G. Nevill-Manning ,T.D. Wu and D.L. Brutlag, "Highly specific protein sequence motifs for genome analysis," *National Academy of Sciences of the USA*, vol. 95, no. 11, pp. 5865-5871, May 1998.

[21] S.F. Altschul, T.L. Madden,A.A. Schäffer,J. Zhang J,Z. Zhang,W. Miller and D.J. Lipman., "Gapped BLAST and PSI-BLAST: a new generation of protein database search programs," *Nuclic Acids Researach*, vol. 25, no. 17, pp. 3389-3402, September 1997.

[22] R. Bhadra ,S. Sandhya ,K.R. Abhinandan ,S. Chakrabarti ,R. Sowdhamini ,and N. Srinivasan, "Cascade PSI-BLAST web server: a remote homology search tool for relating protein domains," *Nucleic Acids Research*, vol. 34, no. Web Server Issue, pp. 143-146, July 2006.

[23] S.R. Eddy, "Profile hidden Markov models," *Bioinformatics*, vol. 14, no. 9, pp. 755-763, 1998.

[24] M. Madera and J. Gough, "A comparison of profile hidden Markov model procedures for remote homology detection," *Nucleic Acids Research*, vol. 30, no. 19, pp. 4321-4328, October 2002.

[25] K. Karplus ,C. Barrett and R. Hughey, "Hidden Markov models for detecting remote protein homologies," *Bioinformatics*, vol. 14, no. 10, pp. 846-856, 1998.

[26] S.R. Eddy, "A probabilistic model of local sequence alignment that simplifies statistical significance estimation," *PLOS Computational Biology*, vol. 4, no. 5, p. 1000069, May 2008.

[27] HMMER: biosequence analysis using profile hidden Markov models. [Online]. http://hmmer.janelia.org/

[28] R. Hughey, and A. Krogh, "Hidden Markov models for sequence analysis: extension and analysis of the basic method," *Computer applications in the biosciences: CABIOS*, vol. 12, no. 2, pp. 95-107, May 1996.

[29] (2001, February) Profile Hidden Markov Model Analysis. [Online]. http://www.biology.wustl.edu/gcg/hmmanalysis.html

[30] S.Henikoff and J.G. Henikoff, "Amino Acid Substitution Matrices from Protein Blocks," *Proceeding of the National Academy of Sciences of the USA*, vol. 89, no. 22, pp. 10915-10919, November 1992.

[31] Richard Durbin, *Biological Sequence Analysis:Probabilistic Models of Proteins and Nucleic Acids.*: Cambridge University Press, 1998, ch. 5, pp. 122-124.

[32] Kevin Horan, Christian R Shelton and Thomas Girke, "Predicting conserved protein motifs with Sub-HMMs," *BMC Bioinformatics*, vol. 11, no. 1, p. 205, April 2010.

[33] S. Kullback and R. A. Leibler, "On Information and Sufficiency," *The Annals of Mathematical Statistics*, vol. 22, no. 1, pp. 79-86, 1951.

[34] L. Steven Johnson, Sean R. Eddy and Elon Portugaly, "Hidden Markov model speed heuristic and iterative HMM search procedure," *BMC bioinformatics*, vol. 11, no. 1, p. 431, August 2010.

[35] V. Kunin , B. Chan B, E. Sitbon, C. Lithwick and S. Pietrokovski, "Consistency analysis of similarity between multiple alignments: prediction of protein function and fold structure from analysis of local sequence motifs," *Journal of Molecular Biology*, vol. 307, no. 3, pp. 939-949, March 2001.

[36] R.B. Lyngsø,C.N. Pedersen and H. Nielsen, "Metrics and similarity measures for hidden Markov models," in *International Conference on Intelligent Systems for Molecular Biology*, 1999, pp. 178-186.

[37] B. Morgenstern, W.R. Atchley, K. Hahn and A. Dress, "Segment-based scores for pairwise and multiple sequence alignments," in *international conference on intelligent systems for molecular biology*, 1998, pp. 115-121.

[38] Gerard Salton and Christopher Buckley, "Term-weighting approaches in automatic text retrieval," *Information Processing and Management: an International Journal*, vol. 24, no. 5, pp. 513-523, 1988.

[39] A.G. Murzin, S.E. Brenner,T. Hubbard and C. Chothia, "SCOP: a structural classification of proteins database for the investigation of sequences and structures," *Journal of Molecular Biology*, vol. 247, no. 4, pp. 536-540, April 1995.

[40] Alexander Chan. An Analysis of Pairwise Sequence Alignment Algorithm Complexities:Needleman-Wunsch, Smith-Waterman, FASTA, BLAST and Gapped BLAST. [Online]. http://biochem218.stanford.edu/Projects%202004/Chan.pdf

[41] Structural Classification of Proteins (SCOP releases). [Online]. http://scop.mrc-lmb.cam.ac.uk/scop/index_prevrel.html

[42] J. Weston, C. Leslie, E. Ie, D. Zhou, A. Elisseeff, and, "Semi-supervised protein classification using," *Bioinformatics*, vol. 21, no. 15, pp. 3241-3247, 2005.

[43] R. Kuang, E. Ie, K. Wang, K. Wang, M. Siddiqi, Y. Freund, and C. S. Leslie, "Profile-based string kernels for remote homology detection and motif extraction," *Journal of Bioinformatics and Computational Biology*, vol. 3, no. 3, pp. 527-50, June 2005.

[44] Y. Yang, E. Tantoso, and K.-B. Li, "Remote protein homology detection using recurrence quantification analysis and amino acid physicochemical properties," *Journal of Theoretical Biology*, vol. 252, no. 1, pp. 145-154, May 2008.

[45] J.M. Chandonia, G. Hon,N.S. Walker,L. Lo Conte,P. Koehl,M. Levitt and S.E. Brenner, "The ASTRAL Compendium in 2004," *Nuclic Acids Research*, vol. 32, no. Database issue, pp. D189-192, January 2004.

[46] Christopher M Florkowski, "Sensitivity, Specificity, Receiver-Operating Characteristic (ROC) Curves and Likelihood Ratios: Communicating the Performance of Diagnostic Tests," *Clinical Biochemical Reviews*, vol. 29, no. Suppl1, pp. S83-S87, August 2008.